ON CREATION AND THE ORIGINS OF LIFE

AN EXPLORATION OF INTELLIGENT DESIGN

BOB YARI

ON CREATION AND THE ORIGINS OF LIFE
AN EXPLORATION OF INTELLIGENT DESIGN

iUniverse books may be ordered through booksellers or by contacting:

iUniverse
1663 Liberty Drive
Bloomington, IN 47403
www.iuniverse.com
844-349-9409

ISBN: 978-1-6632-2610-5 (sc)
ISBN: 978-1-6632-2612-9 (hc)
ISBN: 978-1-6632-2611-2 (e)

Library of Congress Control Number: 2022903485

Print information available on the last page.

iUniverse rev. date: 03/14/2022

For Amanda, Rob, Alex and Andrew.

CONTENTS

INTRODUCTION

THE FOLLOWING IS A SERIES OF THOUGHTS and an exploration of evidence within our current knowledge base to possibly find the true origin of life. Did an intelligent Creator in fact design the foundations of life on earth, particularly human life, with a very deliberate, logic-based design? Or is life the result of a chance-based event, which then grew and advanced with the mechanism of evolution? This writing is a discussion of creation and presents a viewpoint on the possible existence of a Creator entity that is completely unknown to us. This view of creation contemplates the existence of a Creator but one that is unlike the God set forth by most religions. It's a journey from the miracle of cells to various arguments exploring a logic-based intelligence as the Creator of all life on earth.

This exploration of creation is simply one individual's view of our world and our human existence. It does not necessarily hold itself out to be right or wrong. It is meant to present a theory—one point of view—hopefully to stimulate independent thought and debate.

In order to find ultimate truths through evidence-based analysis, the most important feature we must nurture in ourselves is openness and open-mindedness. However much we may believe in our own views and ideologies, we must consider others' different views. We must remain receptive and open-minded in our consideration of other theories, views, and beliefs.

With active thought and open-minded consideration, we can learn to break the constraints and bias of our influenced beliefs, which are often dictated by others and accepted by us without independent analysis and questioning. We can free our thinking from the preconceived beliefs we have accepted from early on in our lives, usually without deliberate and rational filtration—beliefs that often have no foundation in logic and fact. Only then can we truly explore our world and the wonders that abound around us and form educated and rational decisions that we have arrived at purely by and for *ourselves*. This, in and of itself, is one of the great gifts of life.

It is part of the human condition to be compelled to grasp an existing ideology and absorb its beliefs into our own sense of identity. It's a condition that, combined with the fact that we are easily influenced in our formative learning years, makes changing handed-down belief systems a difficult proposition. In this way, religious beliefs, ideologies, and certain prejudices have continued through generations on a steadfast, unalterable course. Despite this propensity, humankind's power of thought and logic gives us the ability to step back from our influenced learning and filter out some of our innate primal instincts. Then we can peruse our world and the multitude of concepts in it through rational and objective assessment. With deliberate effort, we have the ability to evaluate any concept or proposition without preconceptions or prejudices and to find the path to truth through the rule of logic.

In summary, it is my hope that this writing will inspire others to consider various possibilities in our ongoing quest to determine the origin of life on our planet, and possibly in our universe, as we continue our great quest of human learning.

I DON'T KNOW

"I DON'T KNOW." IT'S A SIMPLE PHRASE OUR human mind doesn't like much. It's toxic to human self-importance and self-value. We tend to never use it. Rather, we *know* by accepting or adopting a view, usually someone else's point of view. We listen to the various possibilities and other people's points of view and decide what to make *our own* belief. And we often adopt an entire body of beliefs wholesale, without any rational filtration. Naturally, we are influenced by our mentors, our social environment, and, most dramatically, by the circumstances and atmosphere of our developmental years.

How many people do you know who believe in a religion different from the one they were raised in? Then, how many people do you know who are hard-core believers in the *same* religion that they were raised with? In short, almost all people—extremist believers, devout, or moderate—believe in the religions cast upon them at birth. We believe in these types of environmentally absorbed views with a degree of confidence and certainty that is usually unwarranted. We most often fail to take the question at hand, a proposed ideology, and truly analyze it through careful thought, consideration, and *factual* analysis. In other words, we fail to weigh what we truly know as fact and what we simply don't have enough information on which to render a decision. So, we tend to take a definitive position on issues even though we really shouldn't. At least we shouldn't with excessive confidence and inflexibility. This leads to a common inability to simply say, "I don't know."

We have also come to rely on what we refer to as *gut feeling* regarding what is right and wrong, true or false. This is based on some of our instinctive programming and on what we've come to believe in multiple areas of our lives through our life experience. Gut feelings serve a real purpose. When we don't have facts or enough information to make a reason-based decision, the gut forms an opinion based on everything we've seen and learned to that point in our lives. However, we need to beware of the urge to answer all our questions with our gut. Our cumulative learning and belief systems are often derived without factual analysis. They are usually based on other people's views that have been dictated to us. These are the beliefs that provide much of the foundation of our gut decisions.

Not to say that the gut feeling is always wrong. Sometimes that intuition is the result of learning that is based in truth and actual experience, and therefore it can provide us with beneficial decision-making in certain cases, especially when quick decision-making is required. But since it is also based on beliefs that are unfiltered and adopted at a young age, it is usually highly biased and often incorrect. When we have ample time to evaluate, every question and every issue we ponder should be factually analyzed with complete benefit of the doubt given to opposing views. Only fact-based assessment should drive us to an answer we are sure of when there is plenty of time to contemplate and study.

If after we assess an issue factually and scientifically, we don't have an answer, we should be comfortable saying, "I don't know." This doesn't mean we are not entitled to make choices and to develop beliefs without having everything factually verified. It's certainly normal and acceptable for us to develop unproven views when rational thinking and even some good evidence support conclusions to some level. That has been the basis of many great scientific discoveries. We propose a hypothesis, and then we endeavor to prove or disprove it based on logical and factual analysis. But if there is a lack of fact and knowledge, we should be able to make our assessment and then simply say, "I don't know."

At the very least, we should be able to adopt a position or hypothesis and be very willing and open to modify our thinking in the face of more concrete rationalization or new evidence. Maybe we should simply avoid being so vehemently sure of our position, even if we tend to favor one conclusion over another. By way of example, when the question of God comes up, how often do you hear someone say, "I don't know"? Almost never. But we really *don't* know. Yet we say we are firm believers in God, almost always the God set forth by the religion into which we were born. Or we say that we are atheists and that there is no God. Whether we have taken these conclusions from religious teachings or drawn them from scientific schools of thought (those who theorize life was developed through evolution), some of us are fastidious in our beliefs and defend them vehemently, even though, in truth, we cannot *know* the answer. If we are honest with ourselves, we should be able to say, "I don't know." We can acknowledge that overall truth even if we tend to believe a particular theory or school of thought for varying reasons.

So, to many of the questions and issues addressed and discussed in this writing, my own answer is "I don't know." I certainly have inclinations, and on many, I have convinced myself of a specific conclusion. But I always remain open to new information and healthy debate that can bring truth to the surface. So, where I argue for a point of view, it doesn't mean I have entirely given up on "I don't know."

CONSCIOUSNESS

I THINK, THEREFORE I AM. AS THE DOMINANT intelligent beings on our planet, we certainly know *we* have consciousness. We process information, we think, we plan, we learn, we ponder, we feel. We exist in our minds, and we can interact with and manipulate our world. What about other creatures? Does a dog have consciousness? A whale? An ant? Intuitively, I think everyone believes a dog has consciousness. We can see it in their eyes. They feel, they long, they anticipate, and they even plan in rudimentary ways. Do they have active thoughts? Their intelligence level certainly doesn't allow them to understand higher mathematics. But they can do many of the things our intelligence allows us to do.

Does consciousness require a certain level of intelligence (or vice versa)? I think if a being can perceive its surroundings with awareness, react with its environment, and manipulate that environment with decision-making, it must have *some* level of consciousness. I think the key word is *level*. We have a higher level of consciousness than a dog, and a dog has a higher level of consciousness than an ant. The ant largely responds to its environment with preprogrammed behavior. It makes very limited choices in its life, but it must have a basic sense of itself as it wanders through its environment, robotically doing what it is compelled to do. Do they have internal thoughts? Do they have emotions? Do they have an understanding that they are alive? Can consciousness be correlated to intelligence?

So, if we believe there are varying levels of consciousness below us, are there levels beyond and above us? If we believe in a Creator, intelligent enough to design this life on earth in its various forms, this life that is able to evolve, progress, and continue indefinitely, what level of consciousness does this massively intelligent being hold? How is its awareness different from ours? We can only imagine what that level of awareness in our universe must be like.

What is consciousness? How do we become who we are, observing the world from our unique perspective and interacting with our environments? How is it that we are the one inside our minds looking outward? As the brain initially develops, it is also creating a completely unique set of criteria that becomes the unique individual in a world of billions. These varying criteria must include certain naturally engrained (preprogrammed) personality traits, an intelligence level, and some propensity toward a specific emotional makeup. Then there are the effects our brains collect from our environmental realities as we grow. And maybe there is something external that powers the individual consciousness, like a soul. Or maybe not.

Each of our brains is unique in its composition of all its remarkable components. That must be what creates our individual consciousness. It can't be the other genetic traits that make us up, like physical traits. Our noses, our color, and possibly even our sex may not be material to who resides in our heads, so to speak. We can still be this same person and have a multitude of physical character traits. So, if not for a soul, it truly must be that creation has provided for the brain to randomly configure between a variability that must number in the many trillions, to create a unique individual with a unique consciousness and identity. We are certainly then further shaped and formed as people by our life experiences and environments. But we must start with this very unique iteration of a physical human brain that is uniquely the individual that is us.

The questions of our universe are wide and profound. We are only now beginning to scratch the surface of going beyond our own perceptional limitations as human beings. We, as an intelligent race, have a very narrow set of tools to perceive and to interact with our universe. We see only with the limited visible light spectrum, we hear only waves created through air, we smell limited molecular emittances, we taste within a certain range, and we finally can sense through touch. This touch allows us to conform the "real" nature of the state of matter around us to the picture the brain creates in our minds of what we *think* is our surroundings. In actuality, our brain creates a partial reality for us by creating a picture from the limited available data it has access to. This data comes primarily from electrical signals sent from the eyes to the brain. The eyes don't send a picture. The brain receives electrical signals and, through a miraculous internal process, interprets them into a mental three-dimensional picture. This picture we have is only one version, or interpretation, of the reality around us.

Why are our dreams so real sometimes? Why do certain mental illnesses like psychosis create such realistic delusional worlds for some? Because the brain is using its same ability to create a picture world that it uses while we're fully awake and perceiving our "real" world. This is the world we humans experience, live in, and share. We know no other reality. However, with the advent of infrared imagery, spectral photography, and other systems, we have been able to get a small insight into how our environment may look to a being with different perception organs. Our environment and world would look vastly different if we were to change our perception organs. Imagine a being that could only see heat variations and not detect the visible light spectrum. The picture created in their brain would show a vastly different world.

So, this picture we live in and share with most of our fellow animals is a mental picture based solely on the available perceptive senses we possess. We now know there are many other realities all around us that we cannot perceive biologically. Some of these we became very aware of early on. We cannot see the air, but we can feel its presence and see its effects. We cannot see electricity, but we can see lightning. Other realities exist in our environment that we have no senses to perceive at all, and we must rely on sensors we design and create to perceive them for us. Two of these, nuclear radiation and radio waves or electromagnetic radiation, are now known to us. Although we have no bodily perception tools capable of experiencing these, we now know both flow all around us in great density, primarily from the sun. And, as we have partially mastered their utility, our own modern usage has increased the abundance of these elements all around us without any ability to sense their presence without manufactured instruments.

We now can communicate with instruments we have landed on Mars many millions of miles away, using something we didn't know existed all around and all over us for many millennia. How many other unknown and unperceivable (to us) things exist around us? Our realities are structured by only the things we can perceive. I believe in time we will find more and more of, as of yet, undefined realities that are all around us and in some manner affect our existence and that of the greater universe. These may be things we cannot even conceive of with our current knowledge base. So, again, our picture of our surroundings would look very, very different with various added and subtracted perception abilities.

What is really happening—what the "true" picture of our environment is—we can't currently know. We can only see our partial representation as the picture in our minds, and we can only imagine what we now know *is* there (like various radiations) but that we cannot directly perceive. It's important to acknowledge our natural inabilities. It enables us to think differently. It allows us to move beyond the idea that reality is equal to our reality, that which we can perceive. Once we start believing that, it will greatly widen our horizons of learning and possibilities.

In the continual process of technology freeing us from our biological limits, we are now able to gaze deep into the cosmos and the universe. We have seen new and fascinating things through our limited resource of visible light. With our added ability to detect radiation and electromagnetic energy (using newly devised instruments), we have added an additional layer to what we can perceive from this cosmic gazing. But again, what is out there that we have no bodily senses or invented instrumentation to perceive? What elements we cannot even conceive of fill the universe, bending time and space in a manner we don't yet understand? There could be unknown elements that provide a foundation for other life-forms or intelligences. We know very little of what the universe holds, as we have massive restrictions in our ability to perceive, as proven to us by the relatively new realizations discussed above. We are just scratching the surface. The mysteries of the universe are still way beyond our comprehension. We currently theorize about much of the effects we see with our available instruments. But in reality, we just don't know what's there before our eyes but remains unseen.

I don't know who it's originally attributed to, but this saying goes back to the time of Socrates. Essentially, there are three knowns. We know what we know. We know what we don't know. But we cannot know what we don't know we don't know. If something does not lead to an inconsistency or visible void, we will likely not suspect the gaps in our understanding and knowledge.

And how do our human emotions comprise a component of consciousness? Aside from our information-processing abilities, our brains have been equipped with a large range of emotions. Who we are as individual human beings is highly influenced by our emotional composition. The emotional components of our minds are triggered and activated through chemical releases and absorptions in specialized areas of the brain. Some are need based, like our feeling of hunger, our feeling of lust, or our feeling of love. These are key to assuring the continuation of the species, as they relate directly to maintaining life and reproducing life. Other emotions are environmentally stimulated. They are triggered by external realities. Anger, hatred, ambition, and curiosity represent just a few. These all serve their own purposes in compelling humankind to both advance itself and continue the species. Emotions are part of our internal programming. They are part of our human condition. In fact, some of our feelings and emotions are purely human and appear in no other life-forms on earth. Our emotions are intertwined with the processes of our thinking mind in a chemical symphony within our brains. Although we can now identify areas of the brain that respond to emotional stimuli through brain scans, we know very little about how our brains process and create emotions. Are emotions part of a higher consciousness? They certainly are a critical part of our self-awareness and thought processes. We are, in essence, emotional beings, and our various levels of emotional reactivity are an integral part of who we are as individuals.

Among our significant recent advancements in the last century, we have created computers—machines that can intake data and process that data with internal programming we have designed or written. It is much like our own brain can take data and process it. These computers currently can't come close to our ability to mull and ponder, invent and create. But they do have a cursory ability to compare and sort information. As these computers are given more ability to process through more clever programs we create and are given access to even more sources to acquire data, will they develop consciousness? Will they start to think? What if we are able to program emotions into their function? It's certainly possible, unless we believe there is another factor responsible for our consciousness. Like a soul perhaps. We *are* going to find out someday.

As we design more and more sophisticated processing machines, we will eventually find out if we are able to make them think, create, feel emotions, and ponder on their own. Maybe then we will determine whether a soul is involved in our own consciousness, especially if these machines can find consciousness through programming and processing alone. When we do create these machines, if they have these facets leading to consciousness, then reproducing them will possibly create a new race of being. It will still be unable to repair itself, reproduce itself, or physically manipulate its world. But how far behind could that be? We are nowhere near being able to recreate this system of emotions in our computers with electrical currents and binary codes. But will we be able to do so once we have cracked the secrets of the emotional response system and have found out how to emulate nature's four-element data storage and coding system?

I believe we will get there. We are acquiring knowledge at an accelerating pace, and the fully functioning system in its entirety (our own bodies) is clearly there for us to decipher. And then we'll be one step closer to being able to create consciousness within our own creations. Remember, all the information is here in front of us in our own tiny cells, the trillions of bits of data that can show us how to build a continuing life-form *with* consciousness. We just don't have the advancements in our science and knowledge needed to decode and understand it at this point in our evolution.

In the future, advancements in our knowledge of cell reproduction and its programming will allow us to make certain materials (from human skin to food packaging) from organic elements grown for specific purposes. As we slowly unfold the secrets of our cells, we will be able to build more and more sophisticated machines of our own with unimaginable capabilities. If we survive long enough as a species, we may even be able to design our own new life-forms. Hopefully, by such time, our degree of maturity and scientific responsibility will have grown equally sophisticated so we don't jeopardize the intricate natural balances and threaten the entire ecosystem, including our own existence. We currently live in a time where our technology and science are ahead of our collective maturity in being able to responsibly manage our world, such that we live with a constant threat of annihilating ourselves.

CHAPTER 3

LOGIC AND LOGICAL DESIGN

LOGIC IS THE INCREDIBLY POWERFUL TOOL that provides us the ability to take each step required to arrive at scientific discovery. In the world we live in, things don't just happen. Things happen for a reason. There is a logical explanation for almost everything we see in nature. Our acquisition of knowledge over the many decades has given us more and more insight into the reasons things are the way they are.

Without our current level of knowledge, we were only able to explain many of the things we experienced in nature with mysticism and conjecture. Imagine the relatively raw mind of a cave dweller seeing a rainbow for the first time. How could he explain it without conjuring up myth? The simple realization of cause and effect would make this person need an explanation. Without the science to give an adequate explanation, the mind looks for answers elsewhere. When a logical path is not presented, our human minds tend to believe the most fascinating and magical explanation. Hence, throughout history, humankind has created stories and mythology to explain what it could not with logic.

Various religious and cultural beliefs are examples of this broad body of mostly baseless mythology. As we have gained more factual knowledge through science and rational analysis, we have slowly discredited most of the myths of old. As scientific explanation provides more and more answers to the wonderment around us, we rely less on old, foundationless explanations taught throughout prior generations. In short, logic replaces myth as a way to understand our world.

We now understand how we can create great things with logical designs. We have created our own collection of machines, large and small. That has given us a better ability to look at nature's designs and better appreciate the logic incorporated into some of its workings. We have come to understand and appreciate the rule of logic.

We, as humans, are bound by the natural constraints of the structure and function of our physical minds. We are subject to our emotional makeup, instinctive needs, and programmed reactions. The world around us, however, largely functions on the principles of logic and order. The laws of physics and chemistry primarily determine the workings of our earthly world, our very bodies, and a large part of our perceivable universe. As humankind has gained more and more understanding of this logical world, we have relied less and less on mysticism and myth to explain the incredible world around us.

However, in our greatly accelerated acquisition of scientific knowledge over the last two centuries, perhaps we have come to a certain arrogance of knowledge. We collectively seem to believe that we have advanced and gained understanding of our universe far more than is warranted. We often fail to acknowledge how very little we know, even with how far we have come in science. There is just so much more yet to discover and such a great deal more we can learn and understand. Accordingly, we have tried to answer questions of creation and the cosmos by way of our current state of knowledge. We have, woefully inadequately, tried to explain the origin and extent of the universe as well as the circumstances of our own creation. As we further learn about laws governing the universe, we may be able to answer some of these elusive questions, although we may never be able to fully comprehend the expanse of knowledge binding our universe.

Some of our limitations in comprehension have stared us in the face from the beginnings of mathematical reasoning. A prime example of this is the concept of infinity. We cannot explain infinity, nor can we truly understand it. Like a primate who may demonstrate a degree of rudimentary logic yet cannot be expected to understand quantum mechanics because of the physical limitations of its brain, we too are bound by certain limitations we simply cannot surpass. Our greatest minds have not been able to begin to explain these concepts, which must play into the universal answers we seek.

We strive to explain the limits of the universe by suggesting it is still growing. What is it growing into? What's behind and around it? Taken literally, the concept of infinity means we must first traverse one half a given distance and then again one half of the remaining distance, and so on, indefinitely. So, how do we ever get from point A to point B? If the half distance goes on indefinitely, we technically can't reach our destination. It never stops; we can always cut a given distance in two. The point is infinity doesn't work in our practical world. It's a concept beyond our comprehension. Time plays into infinity. It does not stop, like counting cannot stop. Yet every finite thing is made finite by its limit in time.

We simply cannot fathom the limitless. Everything we know has a limit. If we cannot understand that one concept, can we hope to understand our Creator, our purpose, life, creation itself? No one knows, but we like to pretend we do. We need to fully acknowledge our limitations. As much as we've learned, there is so much more we don't know—maybe so much more we *can't* know. This incredible world we live in on our magical planet has so many fascinating secrets, from its microscopic makeup of a cell that supports the entire larger species of the animal kingdom to the intricate balance of all its life-forms and ecological systems—with the entirety of life being adapted to the physical world through the programs and workings of that single cell. In reality, we know very little of what is visible to us today, let alone what further knowledge is out in our universe that we have no current way of perceiving.

We were put on this earth with no knowledge built into our brains. We have instinctive programming, and we have been given our emotional ranges. And most importantly, we were given intelligence and the capacity to rationalize and learn. Through our slow growth in knowledge, we have come to build great things. We have built ships, skyscrapers, and electronic devices, and we have propelled ourselves to the moon and back. All this was done by learning science, using logic, and extracting raw elements from our basic earthen surroundings.

Like our miracle little cell, which has had that technology for millions of years, we have come to learn to take elements from the earth, modify and purify them with scientific processes, and then build useful items and machines through logical design. It has taken us millions of years to get to the point we are at today. And with all our logical inventions and designs (we have put robots on Mars), we can't yet understand many fundamental mechanisms and the overall logical design inherent within our own cells and beings.

Logic is the foundation of everything we've built to date—not faith, not emotions, and not accident. Our emotions and our faith may compel us to move forward, but only logic has served to truly build and advance us to where we are today. Our creativity in building and advancing has been a usage of logic in design. And therefore, it is logic we must apply to answer the questions of our existence and of our universe. We need to learn to extract emotion and our instinctive programming from our quest in this regard. When we look at the logic applied to our own inventions and also look at the scale of logic incorporated into our biological beings, it becomes more and more convincing that we may not be the only conveyors of logic in this universe. There is so much logic and science well beyond our current state of knowledge. We all must now acknowledge the need to move to more purely logical thinking as we continue to grow and advance our human race.

CREATION

WHEN IT COMES TO CREATION, WE ARE highly polarized. It seems there is no middle ground between creationists and evolutionists. The division is typically between the religious camp believing God created humans (and everything else) and the scientifically oriented school of thought based on evolution. And even with the dearth of knowledge in the area of creation and the manner of the start of life on earth, again, you'll rarely hear someone say, "I don't know." The degree of emotion involved in this debate is immense. People are killing and dying for their religious beliefs. And at the very least, people are typically rigidly opposed to other people's views, often with no more rationalization than having been told what to believe by parents, mentors, leaders (those with apparent authority), and peers in their childhood environment. But there is also a different possibility to consider.

For centuries, religion has guided humans' belief about creation. The story is much the same in all the major religions: God created the earth and humankind around five thousand or so years ago.

Without an iota of evidence, people everywhere believe emphatically in what their own religions have taught them. Stories and histories in books of religion all lay out the manner in which God created earth and humankind. As children, we believe what we are told by adults who guide our lives. We are taught about the religions they believe in. We typically then believe in these without questioning any logical shortcomings. After we swallow a concept as children and it is reinforced into adulthood by people we love and respect, we can believe in even the most questionable ideas, deeply and wholeheartedly. Accordingly, millions of people will kill or die in their dedication to their own version of religious views. The degree of hatred and violence directed toward people of differing beliefs has no bound. To some, it is even acceptable to kill innocent children of a different faith. And all who justify violence and hatred in their dedication to their own school of religious thought do so with the assumed support of their own God, even though the God of almost every religion is supposed to be against killing.

As humans with endless questions about our world, we have a deep need to believe. With this reality of our human condition, a majority of people came to believe in the God of religion—primarily the God set forth by each person's childhood religion. We need faith for many reasons. Probably the most profound is our need to feel comfort and security in our own being, with a greater guiding force watching over us. We also need to explain our very existence, where we are going, and what our responsibilities and parameters are. As science and knowledge provide more logical explanations of our surrounding world, our need for raw faith diminishes. However, where there was very limited knowledge, a great need existed to believe in a greater, benevolent being.

Is the brain hardwired for God? Neuroscientist Andrew B. Newberg, a prominent researcher in nuclear medical brain imaging and neurotheology, asserts that it does appear that the brain has "this profound ability to engage in religious and spiritual experiences, and that's part of why we've seen religion and spirituality be a part of human history."

Clinical psychologist Steven Reiss wrote, "What I'm trying to answer is the nature of why people embrace religion and God." His conclusion: "People are attracted to religion because it provides the opportunity to satisfy all their basic desires over and over again."

Science has debunked many religious teachings, such as the age of the earth and the timing of the dawn of humankind. The religious explanations of creation are scientifically and logically flawed and further muddied by unbelievable, self-serving, and agenda-promoting tales. Given the current state of our scientific knowledge, it is truly nonsensical to continue to believe in the majority of these religious writings. Most likely, they have been concocted and written by humans unrelated to any higher being. And they were most likely written during a time when our science was at its infancy.

But new scientific findings haven't fully invalidated everything we've generally come to believe over the past centuries. Although many of the claims in religious teachings are highly suspect, based on our enhanced view and knowledge of our world, the authors of these religious writings, earthly or otherwise, possibly do explain certain things with some degree of credibility. For example, could there still be a God? Could we just have the details wrong? Certainly, that's possible. But for the following discussion, let's not refer to *God*, as that label is firmly embedded in religious views. Let's call a possible massively intelligent being—whether a single entity, a race, or even a conscious existence of another medium unknown to us—let's call such a being the Creator.

This potential Creator would be a highly intelligent and logical being or force, something undefined other than it possesses immense intelligence, logic, and abilities far, far beyond our own. It can be one or more, it can be matter or nonmatter, it can be of any form, and it can have any purpose in creating life. Much of what it is or isn't, we just don't and can't know with our current knowledge base. But it most likely would *not* be the God of religion. It is not necessarily benevolent, judgmental, or watching over humanity. We cannot know what it is, why it is, or if it is even still in existence, if it has ever existed.

If such an entity exists and if it created nature's entire blueprint of life, it seems highly unlikely that this Creator holds much resemblance to the God of religion. After we use logical and rational analysis without emotional influence, this becomes relatively clear. The God of religion listens to prayers, judges humans, and conveys moral teachings that are to be obeyed at the risk of its wrath. Does God listen to billions of voices in prayer? Does God help some people and not others? Will it let a completely innocent child die of a cruel disease while preaching miracle healing and answering others' prayers? Does this God sit in judgment of humans or are humans purely subjected to the laws of nature, such as survival of the fittest?

When we look at the realities of the natural world surrounding humanity, however created, we see a world without much inherent compassion. We see a world that is largely based on the pure, harsh kill-or-be-killed, survival-oriented laws of nature. This is part of the highly intricate balance in nature where almost every species depends on others for survival, and the balance is maintained not by compassion but by subsisting on fellow species. In fact, almost every sentient living species needs to eliminate another living being to survive. Did a compassionate, religious God create this world of violence and death? Or is a possible Creator not really that interested in compassion and more interested in having the strongest survive to carry on its creations?

It would seem that in the great scheme of life, the individual really doesn't matter at all and is entirely dispensable. In fact, entire populations are dispensable. When we look at the realities of our natural world, what seems to primarily matter in this realm is the survival of the species and not much else. Compassion is largely a human emotion. Some other intelligent species exhibit this quality, but it is rare in nature. And even in humans and these other rare species, compassion is largely conditional. Compassion is commensurate to our own comfort. Simply put, there is no compassion in a starving being, human or beast.

Sifting through all the natural realities of our world and existence on earth, if there is a Creator, it is most likely not the one we've been led to believe in by the main religions of the world. It just may be that this new rational Creator has provided an environment where, as we become more civilized and learned, we will gravitate toward kindness and compassion, not otherwise. This makes sense in a purely rational assessment. The natural order of things is such as to assure the survival of the species. Compassion is a luxury reserved for when survival is generally assured. It would seem that life is an uncompassionate existence with harsh realities, where the individual doesn't count for much until the survival as a species is largely secured. And then compassion and the value of the individual life increases.

EVOLUTION

LET'S FIRST EXPLORE THE VIEW THAT NO Creator was involved in the commencement of life on earth, or the evolutionist view. This school of thought is followed mostly by people with a higher level of education who have rightfully questioned religious norms and beliefs. They have found the God of religion to be fictional and man-made, based on newfound science. Is it possibly a God that was thought into existence to satiate our need for answers to the ultimate questions of how we came to be within the wonder of our complex existence? And ultimately as a tool for humankind to control humankind? Was this a needed authority to control and facilitate human cohabitation without absolute chaos as civilization started to unfold? And ultimately, was it a tool in furtherance of humans' greed for possessions and power? Of course, there is no direct proof that such a religious God *does not* exist. But taken with the totality of claims in religious thinking, some of which *have* been clearly disproven, almost all of its teachings become highly suspect. Thus, the evolutionist view gains further support as more and more educated people logically analyze and then reject the credibility of many religious teachings, rightfully so.

To the American experimental psychologist Justin Barrett, it's clear that humans are predisposed, almost wired, to create religion. Barrett, former director of the Centre for Anthropology and Mind and the Institute for Cognitive and Evolutionary Anthropology at Oxford University, wrote that "our basic cognitive equipment biases us toward certain kinds of thinking and leads to thinking about a pre-life, an afterlife, gods, invisible beings that are doing things—themes common to most of the world's religions."

And the late Joseph Campbell, author of *The Hero with a Thousand Faces* and an authority on mythology and religion, saw each of the world's many religions as a kind of mask reflecting culturally shaped but universally held truths.

Richard Dawkins, the noted English ethologist and evolutionary biologist, rejects the value of religion, regardless of its origins or its influence on the concepts of intelligent design and creationism. In *The Selfish Gene*, Dawkins wrote, "Although atheism might have been logically tenable before Darwin, Darwin made it possible to be an intellectually fulfilled atheist."

And yet the evolutionists themselves have no *proven* alternative theory of creation. The unproven idea is that there is no intelligent Creator and that life happened as a function of earthly chemical interactions that accidentally formed the basis for life. Keep in mind that the evolutionary theory does not contend that complex animals are the direct result of an accidental formation of their highly advanced cells. The concept is that the accidental formation was of the most basic, simple life-form, which then evolved through millions of years into the final life-forms walking the earth today. This idea is difficult to fathom given the unbelievably immense complexity of life, even in its most basic cellular form. In addition to considering the cell's multitude of intricate functions, there is also the issue of the highly complex interdependencies of all species in the balanced natural world.

To address the obvious statistical probability issues of this theory, its proponents often use the billions-of-years argument. And it's not a bad one. The idea is that, however infinitesimally small the probability of an event occurring may be, over the billions of years of earth's existence (an unfathomable amount of time), it can and will happen. It's a hard position to refute, as it stands to reason that nothing is *entirely* impossible, only statistically nearing zero. Accordingly, the theory goes, anything can happen by accident due to random, chance occurrence, given enough time. The mechanics of earth and its elements as they interact with the sun's energy create storms, lightning, and upheaval of chemicals around the earth's surface, all movement and opportunity for any potential interaction to occur. We'll examine this later in detail.

The evidence many evolutionists point to in support of their theory is the biological mechanism of evolution present in all life-forms, the genetic program built into the cell's blueprint of life that shifts life-forms to adapt to their environment in a bid to better survive. Evolution is real; there is no doubt about it. The scientific evidence is powerfully present in our fossil history of life and our newfound ability to determine age through nuclear decay. The evolutionary mechanism is an amazingly complex system whose function we still do not fully comprehend. It involves the ability of a single cell to assess external conditions affecting it and to modify its existential blueprint to adapt to changing conditions. In other words, it entails the ability of the cell to in effect continuously redesign itself. So, the theory would seem to expand that, given the function of evolution, life could have begun *unplanned* as some basic version of an individual cell, the building block of all life-forms. It then presumably slowly evolved into the many life-forms on the planet, including humanity, through the evolutionary mechanism. Again, this is a compelling theory given the billions of years we have had for our accident to happen and the reality of the evolutionary mechanism that could then evolve the basic life-form to more complex beings.

However, in considering this, we must also acknowledge another highly logical possibility: the evolutionary mechanism also has the potential to exist within the structure of life as an element of a *planned* design by an intelligent Creator. If a Creator with an immense knowledge base (many, many times greater than our own limited knowledge) created life (the cell itself and the highly complex organisms it constitutes), it stands to reason that it could have built a mechanism or programming within the life-form for it to evolve and adapt itself to the changing realities of its environment. This would make endless sense in order for that life-form to have a better chance of surviving the constant threat of extinction. If a Creator was to plan life as we know it and plan not to revisit its design repeatedly over time, it would logically incorporate this evolutionary mechanism into its creation.

In any event, the evolutionary mechanism cannot account for the commencement of life itself, as it only exists within the cell structure of life in *full operational form*. It can only be made to possibly account for the movement from a single cell structure to more complex life-forms. In other words, the highly complex evolutionary mechanism would have to have been integrated into that first functional cell, immediately upon its accidental formation, for the rest of the theory to work. So, the accident would not, in and of itself, have anything to do with this feature by way of cause.

In addition to having this evolutionary mechanism within the initial accidental cell, there are many other features necessary for the continuity of the cell's life-form that would have to have been incorporated in that very first iteration of this theoretical cell. However, as far as the evolutionary mechanism itself, it is most certain that this feature would have had to start immediately upon the formation of the first cell. Otherwise, what mechanism would cause it to develop this system over time? It seems it would be a circular argument (like which came first, the chicken or the egg). We need the evolutionary mechanism to develop an evolutionary mechanism. It can't develop over time and many generations of cell modification if it doesn't already exist at inception. In other words, a cell without an evolutionary mechanism cannot adapt and change itself, developing new features. For the theory to work, the incredibly complex nature of the evolutionary mechanism would have had to have been created at the very foundation of this new accidental cell. And this cell would then had to have been the foundation of all other life-forms on earth.

Some scientists are skeptical about the likelihood of such a complex accident. Former geophysicist and Professor Stephen C. Meyer, director of the Discovery Institute, put it this way: "No theory of undirected chemical evolution has explained the origin of the digital information needed to build the first living cell. Why? There is simply too much information in the cell to be explained by chance alone. And the information in DNA has also been shown to defy explanation by reference to the laws of chemistry."

Additionally, for the evolutionary theory of creation to possibly work, we have to address the notion that the very first cell also needed a mechanism to *continue* its life-form over time, once it was initially formed. This is highly likely to be true since the bare natural world we know on earth without life-forms is a degenerative one, breaking down anything that forms by design or otherwise. This may not be true for the greater universal laws that guide all that is in existence (of which we know very little), but it is true for the realities here on earth. Especially with respect to our chemical, carbon-based life, oxidization and other chemical reactions will break down a functioning life-form over time. So, our accidental cell would last a single lifetime and then perish without a trace unless it could readily *reproduce itself.* Therefore, on top of its highly complicated life functions, our first cell would need many other fully functional factors at its initial birth—functions such as the ability to reproduce itself in order to have any chance at all to survive and flourish as a species of life.

CHAPTER 6

THE MIRACLE CELL

WE, AS A RACE, HAVE NEVER KNOWN greater intelligence than our own. All other life on earth is far less intelligent than the human species. The most accomplished and intelligent of us still functions within the boundaries of what we deem as human intelligence. So, inherently, it is difficult for us to conceive of a much greater intelligence. We have tried in fiction. But interestingly enough, even our depictions of God in various media still seems to function at a human intelligence level, complete with human emotions such as anger and vengeance.

In our recorded history, there has been no evidence of higher intelligence. Even our purported representatives of God here on earth have never exhibited a higher level than we have come to know as the normal range of human intelligence. So how can there be another being that *dwarfs* our own intelligence? It's hard to fathom. It is even harder to logically describe such a being. Where is this entity? As we have now peered deep into the universe, where does it reside? Why is there nothing else out there in the cosmos that we have been able to identify with *any* level of intelligence? Why have we not found even a single example of any intelligent design? So, it would seem we are alone in this universe with our maximum intelligence. But *if* we can dispel the evolutionary theory *and* we truly look at the highly advanced science and knowledge built into life on our planet, there truly can be no other explanation than a greater intelligence is responsible for life, an intelligence far, far beyond our own level of comprehension.

Consider that there are more than *thirty trillion* individual cells in the human body, all working in unison to effectuate human life! And the brain contains more than one hundred billion neurons, all interconnected with trillions of synapses—all working in concert to effectuate human thought and function. Is it possible to credit these functioning systems with random occurrence? Nevertheless, to fully embrace the possibility of a Creator, we must first give the full benefit of the doubt to the evolutionary theory of creation.

The idea that there can be a Creator completely outside the religious God is definitely a viable alternative to the two predominating theories. Let's keep an open mind to that possibility as we explore the start of life. I don't know of anyone who has successfully established a step-by-step analysis of how life could have been created and then developed into complex life-forms by this accidental-evolutionary model. So, let's try to analyze how this could have occurred. But first, let's try to define what we mean by "life" in our universe. A simple definition could be this: *a functioning, organized structure that can sustain itself with energy from its environment and is able to completely and continually reproduce its own structure and function to a new separate structure in order to continue its form and function indefinitely.*

"A functioning, organized structure" is basically a machine. Anything we have created that has a logical process and can perform any kind of even rudimentary operation or function is a machine. An old mechanical blade lawnmower, a car, a bicycle, and even electronic circuit boards are examples of machines. As we are progressing in our science, more and more of our machines are functioning with electric current and the movement of information, but they are still machines even as they become smaller and more sophisticated.

A biological cell, within this definition, is also a machine. However, it is built with building blocks and technologies that we, as of yet, cannot build with and do not fully understand. Cells are primarily a chemical-based machine (as opposed to electrical, which is the basis for the most sophisticated machines we have created). But the cell still operates and functions pursuant to a design. It has an energy-driven engine (its mitochondria). It takes in earthly elements that it converts into energy (metabolism), which it then uses to induce chemical reactions to, among other things, transform other earthly elements into biological building materials. It also expels its waste generated by the chemical reactions and transformations, much like an automobile. The cell is a miniature miracle of engineering, whether by accident or otherwise. The very simple function described above is only an iota of a cell's design and capability. Arguably, its most incredible ability is to simply process earthly elements, which basically comprise the dirt and soil on earth's surface, into flesh, bone, and all other bodily carbon-based components. In other words, the cell takes nutrients out of the earth and transforms them into biological materials (primarily protein building blocks) to build additional cells.

In an automobile, we operate as the central brain to control its functions. We have built interfaces to allow us to convey our will into the actions of the machine. A cell operates independent of any external influence or intelligence. Its brain, the element or programming that controls its entire function, is all held internally (in its nucleus). This blueprint information and programming is held in coded format, much like our own computer programs and codes, except *light-years* ahead in its methodology. This internal brain holds the code and programs that dictate the cell's function as well as the entire library of information that forms the blueprint of its design and function. In cells that constitute complex animal life-forms, this equates to an amount of information that exceeds what we hold today in the *entire* Library of Congress—an unimaginable quantity of data.

As sophisticated as all that may be within a tiny cell, it would not survive very long in our degenerative earthly environment. To give a perspective on *tiny*, if you expanded a human body to the size of the entire earth (circumference, twenty-five thousand miles), a cell would be slightly larger than a basketball. "Degenerative," as this environment on earth slowly breaks down everything to its core elements, primarily through oxidization. So, beyond all its functional abilities, this microscopic machine needs a way to continue through time while carrying forward its complete library of data that guides its function. In other words, this miraculous cell needs a way to avoid discontinuation or extinction through degradation. It can accomplish this in only two ways. Either it has to be able to constantly repair and rebuild itself, countering any degradation, or it must create a brand-new duplicate version of itself prior to the original decaying to the point of nonfunctionality. This new duplicate version must also carry forward all its blueprint code, *including* the code that gives it the ability to reproduce itself.

Obviously, we know the cell can reproduce itself. This remarkable ability is a complete prerequisite to life and the continuation of an existence going forward. Our cells, the building block of all life on earth, simple and complex, animal, plant, or other, can duplicate themselves into a new cell that also carries the entire blueprint of its form and function. Imagine if an automobile could do that. Or any machine that humankind has invented. What that little cell can do is light-years ahead of any knowledge or technology we possess.

As an intelligent species, we have made logarithmic leaps and bounds in our knowledge over the last two centuries. Just look at the technology all around us today—machines, all getting smaller and smaller and all beginning to operate with programs, reducing our need to serve as the functional brain of our inventions. Now compare our most sophisticated computers, electronics, and other complex inventions to the little cell. We are not even a small fraction of the way to acquiring the knowledge required to invent a similar mechanism, even when it's sitting right in front of us in its complete functional form, reproducing before our very eyes, for us to *copy*.

In very recent history, we have learned to create computer programs. This began with the basic use of logic gates to filter information, which was transformed into our electricity-based binary coding system. These functions are similar to how our brain processes information. Electric traffic in our brain is carried between neurons by axons and dendrites triggered by various chemicals like calcium and sodium and modulated by neurotransmitters and hormones. The neurons either accept bursts of electricity or they can reject them (presumably a logic gate as well), which gives rise to our thoughts and memories. It is a biological system so complex and advanced that we aren't even close to understanding its intricate functioning.

In our own world of discovering programming, over generations of basic programmers, we started bundling mini programs that then became available to other generations of programmers to build upon. And now we have extremely complex programs as a result of multiple generations built and bundled, one upon the other. However, our most sophisticated and complex programs pale in comparison to the *chemical* programs contained in each and every cell in the living world. We'll discuss this very magnificent chemical programming and data storage later.

Our new ability to create programs that control the function of our machines has given us a whole new understanding and appreciation of how complex a range of programs are needed to operate the functions of the cell. We can now appreciate the degree of intelligent design incorporated into the cell's programming. Take a smartphone as an example. It is a machine. It has capabilities to interface with its human operator, and it can execute commands provided by programming, otherwise known as its operating system or applications. Without programming, the functionality of the phone would be severely limited. It wouldn't be so smart. With early cell phones, we could only make calls. But as we have incorporated more and more sophisticated programs within the phones, their function and utility have exploded, so much so that the machine is now dwarfed by the programming incorporated into it as a ratio of functionality.

This same ratio of functionality is also heavily programming weighted in our little cell as well. As amazing as its mechanical functions are, such as its ability to gather energy and reform earthly elements into new cells, the true miracle of that cell is its programming, which directs its function. We have only begun to understand how this programming is held within the cell. We have found genes and DNA, and we know they are part of an, as of yet, unraveled blueprint of life. But we don't know how that cell is programmed to do everything it does. How does it read its instructions? How does it determine, as it's splitting and recreating itself *trillions of times*, what kind of cell to become in a body as it's building that body *from within*? The programming built into our little cell is light-years beyond the binary code programming we have slowly mastered. It is built on chemical triggers that are based on four or more individual coding elements (DNA and RNA use slightly variant chemical codes), as opposed to our two-element binary code. This gives it exponentially more programming power within its confines.

The programming of the cell is separate and distinct from the data it contains. The programming reads the data to then initiate the processes required to accomplish its many functions. So, not only does the cell carry the blueprint data and information necessary for life and the building of living beings, it also carries a highly sophisticated program that reads the data and then causes actions and processes within the cell to take place. This includes an ability to keep track of the number of reproductions having occurred previously and an ability to track time for its time-based actions.

In a human cell, this programming (which is carried in each and every reproduced cell) is both the data required to construct the entire human body and the programming of how to grow it *from the inside out.* It contains this among many other programs involving the functioning of its components, organs, and life systems. Cells, responding to this programming and using their stored data, start reproducing at great speed and shifting their individual functionality pursuant to their master blueprint to build a bigger, more complex machine (such as the human body). Every cell and function of our brain (including all of its own behavioral programming), all other organs, and their symbiotic operations are contained in that database and programming of the individual human cell. Again, the entirety of this information and programming for *all* its form and function is carried in each and every single cell in the body. So, the single cell's enormous complexity and logical functionality is a vastly more sophisticated machine than anything humanity has ever conceived.

CHAPTER 7

THE VERY FIRST CELL

LET'S EXAMINE HOW LIFE *COULD HAVE* started. To give the evolutionist theory a fair shake, we need to start with its premise of an accidental or natural formation of the basic cell. It is clear that without the cell, no other life-form could exist or evolve. It is the foundation of all life as the core machine that makes all other life possible. So, we have to start with a single cell. In some part of the roiling reactions constantly taking place on earth, energy, movement, chemical, and electrical reactions must have combined in the right alignment that resulted in a complete, functional, living cell, however primitive and rudimentary.

When I say a *complete, functional, living cell*, I mean that cell must have jumped into existence with several key abilities fully intact. Regardless of how basic a design that original cell obtained in this accidental scenario, as previously discussed, it had to have certain elements of functionality in order to survive and propagate going forward into the future to qualify for our definition of life.

There are four basic requirements (listed below) that have to be present at inception. Otherwise, there is no rational path to acquire these over generations, as is required by the evolutionary theory. So far, we have discussed two: the evolutionary mechanism and the ability to reproduce. Without these basic requirements, a fully functioning, most basic accidental cell would simply exist for one generation (the amount of time it could survive on its own) and then go extinct.

This is possibly the greatest dilemma faced by the evolutionary theory. In order for evolution to be able to implement design modifications to an accidental cell, that cell must have the ability to reproduce itself over a multitude of generations so it can effectuate change from one generation to the next. As such, evolution cannot function unless the reproductive ability of the cell or life-form is already actively in place.

Accordingly, this first happenchance iteration of an accidental cell must have, at the very basic minimum,

- been a complete functioning machine with the ability to derive and process energy and elements from its environment;
- contained the complete blueprint of its own design and function, by way of data and programming;
- been able to reproduce an exact replica of itself, including reproducing the blueprint described above on to its next generation; and
- have within its programming the functioning elements required to evolve, either by constantly delivering mutations or by responding to environmental realities with beneficial modifications to its own design (discussed hereafter).

The notion that the blueprint of life, both the data and programming in a cell, would have to be present in the first iteration is supported by the fact that, in all complex multicell life-forms, this information is an absolute requirement for the development from a single cell into the complex being. This information is required to guide the multiplication and specialization of cells within a single generation as it grows into a fully functional being. In this process, each cell must carry *all* information of its development into a larger life-form, as it has no communication with any central command center that carries the blueprint. And with this end requirement, there is no way for a "blueprint-less" cell to later learn or find its own existential data. Therefore, the ability to carry the data and programming (with a specific coding methodology) must have been built into the very first cell.

Logic would dictate that all four of these functionalities must have been present at once for life to commence. Think about it. If a cell formed, and it was this miracle of functionality but did not have the ability to reproduce *and* reproduce its blueprint into the new cell, it would not have the benefit of the billions-of-years argument to evolve. Without the benefit of being able to reproduce itself, it would die in a rather short period of time in a degenerative, oxidizing world. The very first generation would be the last.

So, when we look at the odds of this occurring, it seems that a very, very complex and sophisticated starting point would be necessary, and it starts looking highly unlikely to be the result of a random accident. But in the billions-of-years argument, we can't yet dismiss the possibility that this initial cell of life came into existence as an accident, even as a complete cell with reproductive abilities. I've tried to think of a simpler mechanism that could have been brought into existence and still provide the foundation of life as we know it.

I just don't see how anything short of what is described above, with the four functions being created simultaneously, could lead to the next steps required in the logical path for evolution to build complex animal and plant life-forms.

Despite the discovery of living cells in 1665 by Robert Hooke, it took biologists centuries to learn much more about how cells function and replicate themselves. During the nineteenth century, biologists were able to use the newly invented microscope to study previously invisible components of cells. By the 1880s, scientists had identified the role of chromosomes in heredity, but it wasn't until after World War II that the true nature of the cellular blueprint and replication came into focus.

In 1953, American geneticist and biophysicist James Watson and British biophysicist Francis Crick announced their theory of a double-stranded DNA molecule within a cell that contains all of the information needed to reproduce that cell. As Crick told an audience that year, the team had, in effect, "found the secret of life."

Biologists Arthur L. Koch and Simon Silver saw those required characteristics as essential for Darwinian evolution. "The First Cell," they said, "arose in the previously pre-biotic world with the coming together of several entities that gave a single vesicle the unique chance to carry out three essential and quite different life processes. These were: (a) to copy informational macromolecules, (b) to carry out specific catalytic functions, and (c) to couple energy from the environment into usable chemical forms." One or more of those features probably appeared at some point, they said, but "only when these three occurred together was life jump-started and Darwinian evolution of organisms began."

Before we go on, let's examine parallels to the accident that may have created life on earth. As discussed, given the complexity required of that first life-form, the chances that it happened without intelligent planning and by an occurrence caused by chemicals, electricity, and energy interacting in a random event seem pretty slim. From a human knowledge standpoint, electricity has been the most readily accessible medium for us to discover, understand, and then be able to manipulate into the coding and processing of information. Chemicals, by which life conducts most of its programming and design, are by far a more sophisticated and challenging methodology. In fact, we can't currently even understand how the body and brain utilize chemicals to store data, process data, and communicate within their mass of specialized cells. Despite all our current knowledge, we don't have a clue how the cell's programming methodology works, even when it's functioning before our very eyes. So, where is this going?

It's true that chemicals are readily present on the entire surface of our planet. But also consider that the earth is bombarded by millions of lightning bolts a day. So, now let's apply our billions-of-years argument. Why have we never seen any evidence of an accidental machine based on electric current (with even the simplest functionality)? It may sound silly in a way. But is it any different from suggesting that the unbelievably and profoundly sophisticated cell could have been created by an accident of nature? And as we humans are in early stages of advanced knowledge, we have first found this simpler electricity-based methodology to create our more sophisticated machines. Electricity has been a more viable science for us even though we have extensive history and knowledge of the chemical world around us. It would stand to reason that nature, too, would have easier access to the electrical model for an accidental functional machine to happen.

Let's, for a moment, take a huge leap of faith and say that the complete individual cell somehow did pop into existence by chance (the basic single cell with the capabilities discussed above). We now have to make the immensely complicated transformation into the massive set of animal and plant life on earth today. It still would seem that this amazing little core cell with its incredible coding system way beyond what we can even comprehend today, its ability to reproduce itself, and its ability to transform earthly elements into new living matter would have to have been created with an extremely sophisticated level of planning and design. It is just simply unfathomable that this level of extreme complexity necessary to yield the most basic form of a cell came from an accidental event. Remember, we are not yet taking into account the advancement of this primal cell into more and more complex animals—fish with gills for breathing underwater, birds with feathers for flight, bats that can fly in total darkness using sonar. These are just a few of the millions of transformations that would require immense rational design to function viably. And all these life-forms are able to reproduce using their own unique, intricate reproduction systems, each with a methodology that uses a multitude of interdependent capabilities within each reproductive systems. Can the totality of life on earth be the result of an accident?

Law professor Phillip E. Johnson, often called the father of intelligent design, noted that in Darwin's time, the inner workings of a living cell were impossible to observe. More recently, scientific advances have shown us that "the cell is so enormously complex that it makes a spaceship or a supercomputer look rather low-tech in comparison. So I think the cell is perhaps the biggest hurdle of all for the Darwinists to get over. How do you get the first cell?"

James M. Tour, chemist and nanotechnologist, put it bluntly: "It is time for a temporary time out. Why not admit what we cannot yet explain: the mass transfer of starting materials to the molecules needed for life; the origin of life's code; the combinatorial complexities present in any living system; and the precise non-regular assembly of cellular components?"

THE CELL EVOLVES

LET'S EXAMINE THE PHENOMENON OF A single tiny cell (the human ova) building a fully functioning human body within the span of a nine-month period. Keep in mind that during this entire process, each and every reproduced cell, as it is newly created, needs continuous access to a source of energy, earthly elements (nutrients with which to build new cells), oxygen, and waste removal.

So, initially our foundational cell (the fertilized ovum or the zygote) starts dividing into a multicell clump tightly knit into a spherical mass. As it increases in cell count, it must immediately form a circulation system connected to the uterine wall to provide continuous cellular access to the four main ingredients required for life's continuation. So, many of these initially reproduced embryonic cells will have to have been transformed into specialized cells to form the walls of arteries, veins, capillaries, and an umbilical cord. This happens as the initially duplicated cells (totipotent stem cells) transform themselves into specialized cells as dictated by the blueprint data of the body's design and the program guiding its implementation, based on certain chemical triggers (the genome). This program must keep track of the sequential generation of each multiplication of its cells. This is necessary in order for the program to know exactly what each next duplication of cells (the number of which is increasing exponentially) has to become as a specialized cell (or another more specialized stem cell) and where it must occur in the newly expanding mass that is forming.

Now, our little clump of cells has a circulatory system being supplied by the carrying mother's circulatory system through an umbilical cord. The program can now continue building the multitude of structures and systems needed. Again, this program is carried in each and every cell, which now knows which sequential iteration it is in the entire process. In various areas of our new clump, the program will dictate that certain cells in very specific areas within the clump, trigger a transformation into a specialization, numbering in the tens of thousands of specialties. Heart cells will start building and further develop into specialized cells based on the master plan of muscle cells, arteries, valves, nerves, and many, many other heart cells. Likewise, bone cells will trigger and start further divisions of specialization. Brain cells, eyes, ears, hands, lungs, liver, blood, hair, nail, reproductive organs, bladder, colon (and on and on) cells will start as a single cell and begin their own growth and development path, again following the master plan carried in *each cell*, knowing internally what iteration of the cell reproduction process they happen to be.

Keep in mind, all this is happening without communication with a central command sending out instructions to each cell. This fantastic orchestral synchronization with trillions of operations is taking place with the *entire* plan of the human body being carried by each and every cell, new and old. This totality of information gives the individual cell the information and position within the synchronized programming of the blueprint (by way of knowing its own generational count) to know both its own immediate function as it comes into existence and what its next offspring cell should become. This blueprint tells each individual cell what it needs to execute in order for the body to build from within. In other words, each and every cell must know or carry the information of where it fits in the entirety of this massive build scheme of the human body. It needs to identify its exact timing and function within *trillions* of such individual operations. The miscalculation or misplacement of a single cell could result in a nonviable individual.

All these new specialized cells continue growing and transforming as they slowly start taking on more and more active functions, like the heart starting to pump the circulation as opposed to counting on the mother's heart. In this fashion, each and every organ and bodily system is built internally within the forming fetus body until it is ready to be expelled into the world, living independent of its mother, starting to grow larger and starting to learn. *Mind-boggling*. It is hard to fathom how, even with all our knowledge to date, we could conceive of such a miraculously intricate system.

From our experience in the known universe, machines don't *happen*; they are created by intelligent design. But the design and capabilities of the cell are unfathomable to us humans, even with all the sophisticated machinery and programming we've created to date.

Again, as this brief discussion illustrates, each human cell's library of information dwarfs any quantity of data we have been able to store on our storage devices designed to date. Its programming is light-years beyond anything we've been able to imagine. Its capabilities are confounding. It basically is able to build every life-form known from the basic components of raw earth. Could it have popped into existence in a soup of chemicals by way of an accident? Highly, highly unlikely.

So how does all this data and functional programming fit into a microscopic cell's even smaller nucleus? Remember, the entirety of the information and programming, including all the additional programming incorporated into the brain itself in its final form (with its store of programmed behavior and chemistry to create emotions, just among a few of its complex functions), has to be carried in full by each and every cell in the body, no matter what its function and specialization ends up being. A skin cell in the toe also carries the complete data required to grow a functioning human brain!

We briefly touched on our ability to program and the use of our binary coding system. This system is relatively simple, as our ability to carry coding in electrical current is limited to two varieties of information, either a one or a zero. So, we have created a coding system where all information, numbers and letters of the alphabet, have been assigned a code of sequential ones and zeros. We have learned how to store these sequenced electrical ones and zeros on storage devices in several forms. We have then devised programs that can decode these sequences and commit operations (like comparisons or calculations) and spit out results translated back to the numbers and/or letters that the sequences represent. This is our binary system.

The binary system is very limited, as the flow of information only has two options with which to code its data. What if we could actually use zeros, ones, twos, and threes to encode data? Going from two to four variables allows for an exponential increase in both storage and programming economy, as the code needed for any piece of information is greatly reduced in size. For example, if a given number was electrically coded by using a thousand zeros and ones, that same number could be represented by only a few dozen zeros, ones, twos, and threes. This would result in a tremendous economy of data and programming storage. In fact, we are now creating so much data on a daily basis we are running out of storage space using our binary system. If we were able to find a way to create the functions of coding using a four-element system, our storage capability would increase a thousandfold. Unfortunately, we are far from that goal, even with our great advances in technology and computing power.

However, this technology exists in our microscopic cell! Again, right before our eyes (with great magnification), the cell stores its data and carries its programming with a four-element coding system. It uses a coding system of four separate nucleotides (adenine, guanine, cytosine, and thymine) sequenced together to both store data and write programs to act upon the data. This coding is held in the DNA strands we have discovered. But, again, we cannot yet even decipher this highly, highly sophisticated coding system. We have learned to view it, compare it, and even edit parts of it, but we cannot understand or duplicate its science. Is this incredibly advanced coding system the result of an accident?

Moreover, as we make the move from unicell organisms to more sophisticated multicell organisms, the programming and coding within the nucleus of the cell also would have to expand in complexity and functionality exponentially to accommodate the new specialization in cell functions and their symbiotic relationships. The complexity required in the programming of these new cells is a massive leap forward in their technology. The highly sophisticated coding system built into every living cell allows these new programs to be contained and function within the microscopic cell's nucleus.

Despite the inconceivability, for the time being, let's say that our single cell, with the discussed four functionality requirements, has been created by a confluence of natural events and it is thriving, reproducing, and plentiful on earth. These beings are unicellular organisms (prokaryote). So now, they must evolve into a more complex, multicell organism on their path to the millions of super complex life-forms we know on earth today. The cells in these multicell organisms (eukaryote) are foundationally different and more sophisticated than the cell of the unicell. It is hard to imagine how our single-cell organism can evolve into a multicell organism through evolution given the immense jump in sophistication of the cell programming and functionality. No scientifically substantiated explanation has been presented to date. There are several theories on how this could have happened, and there is a roughly five-hundred-million-year gap in earth's currently known history to account for this change. Some theories claim a unicell absorbed another and became the first eukaryotic cell. Every theory generally relies on some variation of evolutionary mutations occurring within the unicell.

There are two possible evolutionary mechanisms, and they can potentially exist in tandem. The first is a pure response-to-environment mechanism, which edits or amends the organism's blueprint or genes to adapt to environmental realities. We have only recently found evidence (variations to the ends of DNA strands, known as "telomeres") that this type of change can present its effect in as little time as within a single generation. The second mechanism is a random series of mutations in a single generation of the organism, which is then subjected to the natural selection real-world tests to determine if the mutation will persist or die off.

In our unicell to multicell transformation under the evolutionary model, it is most likely that a mutation would be responsible again, relying on the massive scale of time to allow for the unlikely odds. But let's accept the transition on blind faith since there is no factual scientific explanation to date. Let's say that a series of megamutations creates a multicell organism directly from a unicell organism. So, now we have a small clump of cells attached together, acting in tandem as one organism.

This multicell organism is able to remain relatively simple, as each cell is exposed to the exterior environment and can derive its energy, intake earthly elements, access oxygen, and eliminate its waste directly. An example of this type of organism would be common bacteria. Any further adding of cells that are not simply strung together in a line or a relatively small cluster will require complex systems to carry to each cell its requirements of energy, earthly elements, oxygen, and waste removal. These new simple multicell organisms could reproduce themselves by several means, primarily by splitting off and releasing a new clump of cells, known as a blastema, with the complete blueprint to further multiply and recreate its original parent. This class of reproductive process is known as asexual reproduction. This means that a multicell organism recreates itself without a mate, and its cells are primarily identical to the parent cell. This method of reproduction is limited in its evolutionary efficacy. It can evolve very slowly and doesn't have the benefit of bettering itself with the constant shifting of chromosomal traits derived from a mixture of various iterations within the species.

In contrast, with sexual reproduction, each of two required parents provides a sex cell (gametes) that will contribute exactly half of the chromosome base of the parent cell into a joining of two sex cells (one from each parent), which will then create a new organism with traits derived from two varied mating parents of the same species. This allows evolutionary functions to be sped up and allows constantly varying traits in new generations to be implemented into the species through superior functionality and survivability. In other words, the sexually reproduced organism is much better suited to evolve and progress as a species.

We next need to evolve to the very first complex multicell life-form. This new complex life-form would be the very first iteration of life that had complex life support systems and required an entirely new method of reproducing itself due to its inherent complexity. What would this first iteration life-form look like? Certainly, nowhere near a super-complex life-form like mammals. It would most likely still be similar in size to a bacterium. We need to look at possible steps, presented by mutation or environmentally triggered auto gene editing, that would give us a crude first complex form. So maybe it would just be a larger cluster of cells but have developed some form of internal circulation system so all its cells had access to life-sustaining requirements. However it was formed, with complex internal systems, the methodology of reproduction must change drastically to accommodate the duplication of the parent organism. This change must allow the reproduction to have a complex methodology so that the offspring can be introduced into the world with enough systems operating independent of its parent so it can survive earth's environment. It's a process that requires fertilization, gestation, birthing, and postpartum support, among many other processes, until a viable independent version of the species develops from growth.

TRANSITION TO COMPLEX LIFE-FORMS

SO FAR, WE HAVE GIVEN THE BENEFIT OF THE doubt to the evolutionary theory and accepted that within billions of years, however minute, there is a possibility our world is now full of simple multicell species, such as protozoans and bacterium, reproducing asexually. And now we must continue on our path to create a world of complex animals that have developed thousands of intricate and interdependent systems to allow them to function—that is, the complex beings roaming earth today. We need to address the reproductive adaptation from simple cell-splitting of the unicell into cells that constitute the complicated methodology of developing a new offspring in the complex life-forms. This new complex life-form has to develop a new system, still involving cell splitting but one that can arrive at the end point by having its cells morph into various specialty cells—cells that are able to provide the complex systems needed for advanced animal life-forms. These transforming cells are what science is now calling "stem cells," cells that can morph into a multitude of specialized cells. This has to be the methodology, as all life-forms begin by expanding and growing from one single cell. But for now, let's assume the transformation of the cell to stem cells is a reality despite the additional incredible complexity of this new cell. Let's focus instead on two significant interrelated problems that are not solved by an explanation of evolutionary change over hundreds of generations, as would be required for the profound changes needed in the DNA of our new complex animals.

In order to arrive at the advanced and complex life-forms on earth today, there had to have been a transition from asexual to sexual reproduction in the evolutionary path of that first cell. This is a key factor in justifying evolution as the basis of our creation. Animals with complex bodily functions that allow them to survive and manipulate their worlds have to have sexual reproduction in order for these species to survive. This is largely evident, as there are no significantly advanced species (the ones we know as animals, including ourselves) that do not depend on sexual reproduction to continue the species, diversify, and better their kind.

So, the first problem is our transition to sexual reproduction, which requires two parent organisms, from the more basic asexual reproduction of our simpler multicell life-forms. In a world of unicell organisms like bacteria and multicell organisms with asexual reproduction, we presumably now have a planet full of life equipped with evolutionary powers. But we need to transition to the far more complex life-forms that reproduce sexually to create genetic diversity in the species and allow further evolutionary modifications.

How can this happen? How will a cell that reproduces by first splitting its chromosomes into two exact sets and then splitting off into two identical organisms develop or evolve into a cell that offers only half its chromosomes to a new cell so that the new cell can then find another unrelated cell to join together, reconstituting a full set of chromosomes and thereby creating a modified new organism that then becomes unique within its species?

In this realm of giving the total benefit of the doubt to the accidental or mutated development, the real issue becomes this: how can the change occur over time to accommodate our great billions-of-years equalizer (now actually dropped to millions of years)? The problem is that there isn't an intermediary step that is viable, as we are contemplating a major transition in *reproduction* methodology. The fact that we are contemplating a materially significant change in reproductive methodology eliminates our ability to change over time or over several generations. As we are moving from one super-complex reproductive system to an even more complex reproductive system, the change must occur in a single generation and still be a viable reproductive methodology; otherwise, it could not continue its life-form. Given that it is the reproduction mechanism itself that is making a dramatic change in functionality, from asexual to sexual reproduction requiring two independent sex cells, the fully operational changeover has to truly take effect in one generation.

We may hypothesize that maybe there is an intermediary step between these two types of reproduction. But there is no in-between iteration that can possibly deliver the same twenty-four chromosome structure. Either it has to be an exact copy or a mechanism must evolve that allows for a sharing of partial chromosomes from two separate parents in the same species. Think about it. The magnitude of this change is compounded by the *contemporaneous* need for another organism within the species evolving and delivering the identically formed counter sex cell with exactly the same twelve chromosomes. Even if we could bridge the transition with changes over generations to arrive at a sex cell, we would need a wholly independent, separately synchronized evolution of the exact same nature to occur in another member of the same species. So, how are these two separately evolving cases reproducing in the interim?

If we can somehow believe that a mutational change occurs in one generation that can provide this miracle sex cell, it would simply die off if it didn't have an exact mutation of another parent that happened in the exact time frame of that same generation to mate with. This does not even consider the astronomically low odds that they would meet and combine in the expanse of time and space. This would compound the odds in a negative manner to an almost infinite degree. There seems to be no rational explanation to bridge this divide. I have not been able to conceive of any, or find reference to any, intermediary step. In my opinion, a full jump to the final sexual methodology is necessary. And such change must happen in a single generation. It is also such a drastic change, requiring an unimaginable degree of mutation *and* a third-party separately derived mutation (in the same single generation) that is identical and can provide the mate cell, that we are now outside the realm of possible, and we no longer have billions, millions, or even a thousand years to justify the odds. That profound of a change taking place in a single generation and in two completely separate but identical and contemporaneous mutational events is completely implausible.

In *The Masterpiece of Nature: The Evolution of Genetics and Sexuality*, English evolutionary biologist Graham Arthur Charlton Bell called propagation through DNA exchange between organisms a tremendous challenge to evolution theory: "Sex is the queen of problems in evolutionary biology. Perhaps no other natural phenomenon has aroused so much interest; certainly none has sowed as much confusion."

According to David Mark Welch, evolutionary biologist with a background in biochemistry and molecular biology, "We still really don't know the answer to this very most basic question; we don't know why sex exists."

After careful analysis and considering all the possibilities, it seems that even if we are to give the billions-of-years argument all the benefit of the doubt, we still cannot account for the fact that getting from simple cellular reproduction to the very sophisticated and intertwined complex animal sexual reproduction requires massive leaps within a *single* generation.

Reproduction is the key prerequisite to the continuation of a life-form. Therefore, large-scale modifications to its processes require an immediately fully functioning transition by the very next generation. In fact, *any* change or modification has to still result in a fully functioning reproductive system. Otherwise, the life-form would cease with any intermediate form of reproduction that did not make the complete jump to the *viable* next version. Therefore, we only have one generation to make evolutionary adjustments to reproductive systems. And no amount of genetic mutation could ever account for the massive degree of sophistication in the complex animal sexual reproductive system.

Now let's examine a second significant barrier to evolution alone accounting for the ultimate design of advanced life-forms: the challenge of ending up with the super-complex animal life with multiple complex systems required to sustain its life—an organism that is then able to gather its fuel, think, manipulate its environment, have purposed mobility, and reproduce itself among thousands of other capabilities (today's advanced life-forms on earth).

From the very beginning of this transition, there has to develop a whole new methodology of reproduction. We now have to accommodate the growth from a single newly fertilized cell to the final form and functionality of the complex organism. In essence, once an organism has made the transition to sexual reproduction and takes on a complex form containing numerous internal systems, we can't simply drop a single cell on the ground, hope to have it find a corresponding sex cell, and then have it grow into a new complex being. This new methodology of reproduction requires a way to find a suitable second mating parent, an attraction to mate with this second parent, a way to protect its early combined (fertilized) cell, and a way to provide it with nutrition and oxygen, all for a duration of months of growing time. It needs a way to deliver the new being into the world and then nurture it until it can fend for itself. All the systems we see in animal life to allow this new complex reproduction would have to have evolved at once. Each and every part of reproduction, from the male-female attraction to the incubation environment, down to the birth canal delivery process, must come together at once in the very first transition from an organism that can reproduce by creating another parallel cell to one that needs complex biological and environmental designs to create an identical complex life-form.

But the great catch is that, again, this immense transformation can have only one generation to so evolve. There will be no next generation if the male of the species isn't programmed to intensely want to inseminate the female, or if there is no perfectly timed trigger to initiate the labor contractions that are needed to expel the unborn from the womb, or if the ovulation cycle isn't refined to its final form. There are hundreds and hundreds of prerequisites to a successful reproduction that all have to work in tandem. This all may theoretically happen over billions of years but not if there is a generation-stopping event. Everything can evolve, but certain elements of the reproduction system have to be already functioning at optimal levels before we can have one further generation. The essential problem is that much of everything having to do with complex reproduction cannot evolve over time and beyond a single generation to provide instant functionality.

Additionally, to make the transition to complex life-forms, even if we assume the reproductive challenge was somehow met, hundreds of separate body systems must evolve and function in concert with one another. Pulmonary, circulatory, endocrine, muscular, nervous, urinary, digestive, and many other systems have to evolve together. If these changes evolve out of sync with one another, you will have one or some working systems without others being fully formed. As we clearly know with our current state of medical knowledge, an incomplete complex multicell life is unsustainable without *all* its major systems functioning in concert. And without full and synchronized functionality, that life-form would perish. Today, all it takes is one main system to stop functioning properly, and that life is terminated. Hundreds of primary biological systems have to be fully operational for life to be sustained. And in order to get to the complex life-forms, all this evolving of major life systems would have to occur in concert with eyes forming, brains developing, and life-essential subconscious programs being created.

Keep in mind, even developments like sight and certain instinctive programming would be immediately necessary in our new complex multicell organism for it to be able to survive past one generation. Can this all happen in concert with evolutionary changes and mutations? It is a phenomenally complicated and fragile system with an intricate balance of chemistry and function. The most minute divergence from its final form in most any system causes that life to become unsustainable. Accordingly, the rational odds for these occurrences to happen in the required synchronized fashion are minute in the trillions to one, maybe less.

The probability factor aside, even Darwin could not explain how some of the current components of complex life-forms could have resulted from an evolutionary process in their eventual creation (i.e., forming out of necessity or betterment through a step-by-step progression without being logically planned from inception of the life-form). Such components include the seeing eye. Many body parts can evolve from a single cell to their current forms while being fully functional in each step of their developmental cycle.

This developmental path (growth within a generation) is thought by scientists to follow the evolutionary path (change over many generations) of the resulting biological component. For example, the fetus acquires a very small and fully operational heart early on. The heart can function properly from a very small unit and progress or evolve into the final-sized organ, serving a fully grown body. However, some body parts, such as the eye, cannot function in any other form other than their *final* form. In short, in the possible evolution of the eye, there is no step preceding the final form that would have any value to have evolved to that point. If such is the case, evolution cannot account for the eye. The eye would have had to have been planned into the body from its inception as a completed, functioning element. This has challenged evolutionary theorists over the decades.

As Walter Jakob Gehring put it, "The eye was invented only once." Gehring, the Swiss pioneer of molecular developmental biology, wrote, "Finding that a multi-control gene for eye development in mammals was much the same as that in flies suggested that the eye was invented only once. I found out that it has evolved only once and then radiated out into these very different eye types."

Darwin himself found it "absurd in the highest possible degree" that a mutation could deliver such an intricate mechanism: it is difficult to rationalize a step-by-step development of the eyeball since it really doesn't function in anything but its final form. This is the reason babies are born with eyes proportionately larger than their final configuration within their growing heads. To provide vision, the eye must be at full scale. Due to the focal point critical distances present in the internal eyeball, a smaller version or developmental step in its evolution that could grow or develop further will not function. This is squarely contrary to the deemed operation of evolution. Evolution can't *see* beyond its next stage. The evolutionary mechanism effects change that provides a direct benefit, such as nostrils becoming larger due to air becoming thinner in hotter climates. Evolution does not incorporate change into its host's blueprint that it somehow anticipates will bring future benefits with future compounded changes. It cannot foresee or plan the future for further development to provide a benefit to its organism. As another example, how did feathers evolve to allow flight in birds when a very concise configuration and length allows actual flight? This would require logical planning, something only provided by intelligent design. Likewise, a partial mutation of no benefit to enhanced survival or function would not survive past its single generation or wouldn't be incorporated into the next generation of the species if the mutated organism did survive.

As discussed, there are thousands of different biological systems and corresponding cellular programming that need to be working in perfect concert to allow two mating pairs to create an offspring. These systems require a male and a female, each with hundreds of their own unique systems that need to be functional and completely in synchronicity with their very intricate roles in the overall reproductive scheme of their species. But for a male and female to develop these systems over time, assuming they could simply jump into existence by some mutational occurrence, the two separate and complementary systems (male and female) would have to evolve at the same time and result in the complete male, with all functions intact, and a complete female, with all functions intact. Keep in mind that even a single exclusion of a system could result in nonreproduction and therefore termination of the species. For example, imagine if the male and female developed simultaneously and completely separately with their respective reproductive systems all completely viable, but there was no breast milk to nurture the infant of the species right after birth.

There are hundreds of simple exclusions that would cause an extinction. Imagine two primitive but complete humans, one male, one female, newly evolved to be able to mate and reproduce an offspring as humans do today. We can imagine them being smaller, hairier, and more apelike with much smaller brains. But what does their last iteration look like? The one just before the complete functionality of this new reproductive system. If it did not have *all* the systems required for human reproduction, how did it survive and reproduce?

LIFE AND INTELLIGENT DESIGN

WHEN WE LOOK AT THE MAGNIFICENT capabilities of our core biological cell and we consider that the very first iteration had to have four key functionalities (listed earlier) built into it, it is difficult to imagine how it could have been created without logical design incorporated into its creation. We have analyzed the difficulties presented to any theory that starts with an accidental event and uses the evolutionary mechanism to end up with complex life-forms. We have also discussed the enormous complexity of the coding methodology of the programming and data built into each and every living cell on earth, technologies that are so far beyond what we know with all our advancements to date. It's just simply too grand in its scale of science and too balanced in its intricate interdependencies to be void of intelligent design.

Then there is the body's remarkable ability to repair itself. If we adhere to the theory of evolution, it clearly follows that the lower and simpler forms of life proliferated before the complex life-forms appeared on earth. Therefore, an evolving advanced animal form would be subjected to a world filled with bacterium and other microscopic life-forms ready to invade and ultimately destroy its body. An advanced animal is placed in tremendous risk when any of its protective skin develops an opening. Even a scratch exposes the body to invasion. We survive most injuries to our outer skin by virtue of our body's ability to repair itself, in addition to other forms of internal repair of bones and organs. The cells in the afflicted area, without any communication or command from the brain, will begin to speed their reproduction in the exact correct pattern to slowly replace lost cells, expel foreign matter, replace circulatory vessels, and close any external opening of the skin surface. These cells, again, each contain the full data and programming of the entire body, and they know their position and function within the body. Somehow, they learn of an injury and are put into action pursuant to a repair scheme that also must be carried in each and every cell (since the brain does not initiate bodily repairs). The cell then knows exactly how to accelerate its reproduction and to produce the exact type of cell needed and in the exact correct position and sequence to effectuate a repair.

Imagine the state of a body after a short period of life if the ability to repair did not exist. The accumulation of scratches, cuts, and other injuries would have us literally torn apart and quickly subjected to the wrath of infection. No complex life-form would get far into a young life without dying off, especially before our new medical knowledge and treatments. Life would certainly not progress to a point where a complex animal could reproduce. The same is also true for the body's immune system and its ability to fight off external microscopic intruders, viral and bacterial. The question is, how can these repair systems evolve into existence (requiring multiple generations) when they are essential for any given generation of a complex life-form to survive long enough to be able to reproduce? Remember, for any evolving to take place, we need multiple generations. This is another conundrum of passage from simple multicell organisms to complex multisystem advanced life-forms.

When you start adding the layers of complex systems, many of which cannot transition past a single generation if they are not already incorporated into the DNA of the species, it becomes pretty clear that for the evolutionary theory of unplanned and accidental formation of life to be responsible for the creation of life, you need the cell of an entire fully functioning animal (both male and female) to be a part of your accidental start. That starting point is way too complex and requires the contemporaneous start of an opposite sex mating pair. There are just too many major hurdles to rationalize an accidental start of life as a "basic" cell that would then be further developed through the evolutionary mechanism. The science is just too great and profound. The systems are too logical and too perfectly balanced for a human cell to pop into existence. Then you have the entirety of the natural world and its interdependencies. How do the intricate balances in the totality of nature occur as a result of evolving? We live in a world where the loss of a single critical species could result in our own extinction.

The apparent simple truth becomes there is no accidental, mutation-driven, or chance-based explanation for the start of life that is at all viable. And as such, we are left with a single explanation: that the blueprint of complex life must have had intelligent design behind its creation. It had to have been planned and executed much like we plan and build any sophisticated man-made machine in existence today. And we probably can extend that explanation to the creation of the simpler foundational cell itself; life, as we know it, is manifested by intelligent design. And that is a limited conclusion. There is nothing in this conclusion that tells us anything about the who, why, when, and where with respect to a Creator. Even if we accept that life was created by an intelligent entity, we have to acknowledge we know nothing about that entity.

Sir Fred Hoyle, the astronomer who coined the phrase "big bang," believed that "some super-calculating intellect must have designed the properties of the carbon atom, otherwise the chance of my finding such an atom through the blind forces of nature would be utterly minuscule."

Having a large amount of carbon in the universe makes it possible for carbon-based life-forms of any kind to exist. For Hoyle, a commonsense conclusion was "that a superintellect has monkeyed with physics, as well as with chemistry and biology, and that there are no blind forces worth speaking about in nature. The numbers one calculates from the facts seem to me so overwhelming as to put this conclusion almost beyond question."

As real as evolution is, after careful analysis of the stepping-stones of transitioning life from a single cell to complex animals, we are led to believe that evolution has to be a system built into life as an element to provide enhanced survivability to the species and allow an organism to better itself given its environmental realities. It is a mechanism that has served life well and is maybe even the reason life endures. However, the theory that a cell is formed by an accident in nature begins to seem factually impossible as we factor in all the hurdles discussed. Think about the required complexity of that first accidental cell that would necessarily include the reproductive ability (recall that you cannot evolve a reproductive ability since evolution requires multiple generations, which requires a reproductive ability). Think about the additional difficulties of evolving into complex beings, including the complexity and programming of the human brain. Think about the ultrasophisticated coding system used in the blueprint of life. Imagine the level of intelligence and design required to build an entire human body from a single cell. Add all these highly implausible odds together, and even billions of years cannot possibly provide the basis for an accidental creation.

I, and others, believe this can lead to only one conclusion. Life has been created by great intelligence—intelligence way beyond our human level of intelligence and knowledge. By whom? I don't know. I certainly don't think it's the God of religion. That definition of a Creator has now been soundly debunked by science. But I do believe an immeasurable intelligence is responsible for the entirety of the design, function, and interdependencies of all life-forms on earth. In a way, it's very exciting to believe that. We can accept that religion has been an elaborate story and yet still accept that there *is* some great force of knowledge out there whose intelligence is so far beyond what we can even imagine—all for us to find and learn as we expand our knowledge base.

Is this force a being? A field of energy? A form of intelligent life of which we cannot conceive? Is it singular or a collaboration of intelligences? Does it watch its creation? Is it even in existence still? *I don't know*. But I believe we may be able to find out as our knowledge base grows and expands. Maybe that's why we're here. Could we be placed here to see if we can find our Creator? We certainly have the fascination, curiosity, and drive programmed into us. Or maybe the culmination of our quest of knowledge is to create our own life-form and put it on another planet. Can it find us if we give it intelligence, curiosity, and an ability to explore and learn? Even if we provide it with no knowledge from us and no knowledge of us?

In the end, the point is to simply start with a blank slate when we attempt to answer our questions. To erase the blackboard clean and to start populating it with facts and honest analysis. To cleanse our judgment of biases, new and old. Biases taint our conclusions and carry forward barriers to arriving at independent and clear decision-making. We can do this, but it is not easy. It is asking ourselves to discard deeply ingrained ideologies that at times make up a part of our very identity. It takes tremendous strength and internal reflection to be able to accomplish this. But only when we clear our minds of illogical influences can we then use our full intelligence to get closer to the answers we seek. Often, this journey is one where we end up not knowing any answer. But that is OK. It prepares us for fair and proper assessment when more facts present themselves. It is far better than simply accepting someone else's ideology without question. Think, question, analyze, and arrive at your decisions without bias and with skepticism. It's the difference between being the pawn or the player.

Printed in the United States
by Baker & Taylor Publisher Services